另 壹 种 **表达** DIFFERENT EXPRESSION

沈瑾建筑设计作品集

1965年生 / 工学博士 / 高级建筑师 / 国家一级注册建筑师 / 天津大学兼职教授 / 中国建筑学会理事 / 获法国总统奖学金"百名建筑师在法国" / 全国政协委员 / 全国政协"人口、资源与环境委员会"委员 / 唐山市政协副主席 / 唐山城乡规划委员会副主任

沈瑾
SHEN JIN

Born in 1965 / Doctor of Engineering / Senior Architect / National first-grade registered architect / Guest professor of Tianjin University / Councilman of Architectural Society of China / Participated in "one hundred architects in France" project funded by French President Scholarship / Member of The CPPCC National Committee / Member of The CPPCC National Population / Resource and Environment Committee / Vice Chairman of Tangshan CPPCC / Vice Director of Tangshan Urban and Rural Planning Committee

作者简历 .... 另壹种 **表达**

序言 另壹种表达

近年来,建筑的艺术性和与自然相和谐的生态性,越来越受到人们的关注。作为科学与艺术关系最密切的门类,建筑既有艺术的一面,又有别于纯艺术创作,它在艺术的权重和表征手段方面往往因类而异,并与社会经济、社会文明和文化的发展息息相关,因而建筑的艺术性常表现着随类赋采、有所轻重的特点。

美学家们认为艺术本质上是人类的情感意象,对建筑而言,也概莫能外。可以说,审美是人们对建筑观瞻的普遍性要求,例如"美仑美奂"一词就是对丰富的建筑空间和形态产生愉悦之情的赞喻。然而,建筑的艺术性当不只是一般的形式美,由审美联想所引发的诸多情感意象,皆应属之,甚至能达到更高的艺术境界。

三十多年前,我国经济尚处在欠发达时期,囊中羞涩,又百废待兴,建筑创作局限"在可能条件下注意美观"的国策,建筑艺术只能在个别之中有所侧重。改革开放赢得了中国经济的腾飞,特别是近二十年建筑业空前繁荣,几乎成为世界最大的建筑市场。一批新生代的青年建筑师应运而生,他们可以大胆地研讨建筑的艺术,并通过实践创造出许多别开生面的情景建筑。沈瑾就是这个年代涌现的优秀青年建筑师之一,这本书汇集了他在1997-2008十年间的五个代表作品。

沈瑾幸运地赶上了我国全面步入小康社会的新时代,国家兴旺为这一代人提供了施展才华的机遇,建设昌盛为他们开辟了大显身手的"战场",使之在迅速融入中锻炼成熟。沈瑾可称为其中快速成长的建筑师,他大学原学的是土木工程,1994年凭借个人志趣改读了"建筑设计及理论"的研究生,1997年获得硕士学位,1998年就完成了"潘家峪惨案纪念馆"设计。2000年这个设计获得了建设部全国优秀设计一等奖,接着又连续获得第九届国家优秀建筑银质奖和建国60周年中国建筑学会创作大奖。一个初出茅庐的青年学子能以惊人的速度赢得如此的"开门红",实属不易。但他没有自傲,没有止步,2000年起又主持了"井陉矿万人坑纪念馆"的设计,并再次进入建国60周年中国建筑学会创作大奖入围奖的行列。两个设计都是在事件发生的原址,用抽象的手法使人联想到史实中令人发指的惨景,控诉了战争的邪恶和法西斯的残暴。悲愤、震怒的情感意象产生了震撼人心的力量。这两个未按"套路"设计的建筑虽小,却达到了呼唤捍卫人性尊严的大效果。

沈瑾也是一位出色的两栖型建筑师。2001年由职业建筑师升为城市高管后,他把"作"和"管"结合于一身,既从宏观视野提出要求,又以敏锐的专业视角鉴别优劣,在组织与参与城市设计中,让一些建筑师更好地发挥所长为城市增辉。转变职位后,他虽不再直接承担有规模的工程设计,但也不放过具有寻向意义，且小而精的设计创意。出于对城市景观的关照，2004年他设计了两个小房子：其一是在城市公园中营建了看与被看相统一的小建筑，取得双赢的景观效果；其二是在由采煤沉陷地改造的城市生态公园内，为记录这一变迁而建造的"南湖展览馆"，设计用大片实墙陈展了该地原貌，又以大片落地窗透视着现实佳境，让人在触景生情中领悟到人与自然和谐的正确指向。沈瑾还以建筑师和官员的双重身份与责任感，在城市的改造与建新中发挥了能动的作用。唐山这座旧工业城市经过特大震害后，仅有的遗存珍如凤毛麟角，保留城市记忆更有特殊意义。面对因建新而不假思索的拆旧，他奔走各方据理力争，终得在决策者的大力支持下，策划了"唐山规划展览馆"这项以改造遗存与建新重组的群体建设工程，并从城市全局视野主持整体构思和建筑的具体安排，与"都市实践"的王辉等建筑师通力合作，使设计很好地反映了这座城市历时性的文化发展面貌，实现了寓旧于新、面向未来的都市设计策略。这项设计也获得2010年美国《建筑实录》的最佳公共建筑杰出大奖。

熟悉沈瑾的人都知道他非常用功又善于交友，并能从中吸取他人所长。他的进步不外勤奋加悟性，业精于勤，自不待言，高悟性则是和他钟爱摄影艺术所具备的情感冲动和艺术直觉密不可分。2002年留法回国后他曾出版过一本《建筑师的欧洲视隹》摄影集，独特的取景和用光令人刮目相看。艺术的特点是讲求灵性、拒绝相仿、意在出新。老一辈的哲学家曾将人的智慧分为科学家从微观到宏观，从局部到整体探索的量智，和艺术家从宏观到微观，从整体形象的感受和质变的发展过程去探索性智，并认为性智对创造耳目一新的认识方式至关重要。可见肩负有艺术性创作的建筑师多接触各类艺术，对转换视角、激励创新还是很有好处的。

这本书记录了沈瑾跨入建筑师门槛以来十年间的长足进步。如今他正当不惑之年，还有许多个十年的发展空间，希望他在每个十年都不停步，都能做出不同凡响的成果。

黄为隽2012年谷雨于美国

黄为隽：天津大学建筑学院教授 博士生导师

| 序 言 |……·另壹种| **表达** |

另一种
表达

人类通过各种方式进行思想的交流和表达，最直接的方式是语言文字。作为交流工具的语言，语义的清晰、准确表达是语言的基本要求；具有审美要素和思想要素的语言成为文学的表达；审美和思想表达密不可分，文学的表达能产生语言的新意和新的审美。在诗意情怀表达的同时还应有思想要素的叠加。若缺少思想的内涵，文字虽具形式美的一切要素，即使文学情节再夸张，形式再唯美，也不是文学的常态，也只是徒有形式感的一段文字而已。

思想的表述同时赋予作品的形式以美感；萨特和尼采具有深刻意蕴的哲学思想也是通过文学表达的、触人心灵、教化天下的至深哲理往往是以审美的形式表述为载体的，形式需要思想的平衡。戏剧、音乐、绘画、雕塑等其他艺术形式与文学的表达一样，也都是通过形式审美表达思想和情感，只是表达载体、方式和途径不同而已。不同艺术形式的载体都具有外在技巧的形式表象美和思想情感的内涵美的特征。表象的技巧美和艺术的内在感染力是作者与观众对话的媒介。通过解读作品可以感知作者的思维轨迹和想要表达的内容。艺术具有审美效应的同时行文化层面的考量，进而带有更高哲学层面的思考。欣赏形式美的同时对艺术作品思想的见仁见智解读，恰恰是艺术的功效和魅力所在。

| 序 言 | 另壹种 **表达** |

People express their thoughts in a variety of ways. But language is the most direct tool for communication. The basic requirement of language is clearness and accuracy. Language with aesthetical and mental elements turns to be the expression of literature. Aesthetical and mental expressions are inseparable from one another; and literary expression generates new meaning of language and new aesthetics. Language should incorporate mental elements, while presenting poetic feelings. Without mental connotation, no matter how beautiful the lines are, how exaggerate the plot is, it is not the normal state of literature, but only a paragraph of words with an empty shell.

A DIFFERENT EXPRESSION

Mental expression endows the work with an elegance of form. Sartre's and Nietzsche's profound philosophies are also delivered by literatures; touching and influencing theories are usually presented in aesthetic expression. Forms need to be balanced by ideas. Like literal expression, ideas and feelings of drama, music, painting, sculpture and other forms of arts, are all conveyed by formal aesthetics, only different in expression method and way. The embodiments of various art forms share the same characteristics of beauty in both the outer form and inner connotation. Technical beauty in outer form and inner artistic appeal are the media used by the author to communicate with audience. By reading the work, the reader is able to reveal the author's mental process and what the author meant to convey. Art not only brings aesthetic effects, but also embodies cultural value, and moreover, initiates philosophical thinking. Artistic appearance and various expressions of the work connotation are what the functions and charms of art are in essence.

建筑表达

艺术发展史中，建筑的艺术创造曾涵盖雕塑、绘画等其他艺术形式，多种艺术门类在建筑舞台上得以存在和实现。建筑具有实用性的同时也具有艺术表达特质，建筑以自己独特的方式来表达审美和思想。建筑艺术与其他"纯艺术形式"的本质区别是建筑的功能使用特性以及反映人类成就的工程技术物质性特征。建筑形式表达有别于纯造型艺术，建筑建造的逻辑确立了建筑理性原则，即使是建筑非理性内容也是由建筑的理性手段来实现的，用理性方法追求建筑的多样性。建筑基本问题是人对空间的使用与体验、材料的搭接组合、构造的逻辑、建造的质量，以及同基地环境关联。在解决基本问题的前提下，将技术层面上升到具有形式审美层面成为建造的诗学，进而能从思想层面表达时代的文化精神，承载生活中抽象的思辨。将思想理念凝结在建筑中，用建筑来揭示隐藏在建筑深层构造的哲理与美学原理，这也是建筑学追求的永恒主题。

建筑的科学与理性因素构成了建筑学"形而下"层面的问题。思想意识形态层面的感性内容属于建筑更高层面的"形而上"问题。"形而下谓之器，形而上谓之道"。建筑的"道"和"器"的关系应该是相辅相成、互为表里的。

In the history of art, architectural art used to encompass sculpture, painting and other forms of art. Diverse kinds of arts coexist and realize on the stage of architecture. Besides utility, architecture also has artistic expression feature, conveying aesthetics and thoughts in its unique way. The essential differences of architectural art and other fine arts are the functional features and engineering techniques representing human accomplishments. Architectural expression is different from other plastic arts. The logic of architectural construction is applied to establish the rational principles of architecture, even though the irrational contents are realized by rational means. And, the diversity of architecture is pursued with rational method. The fundamental questions of architecture include: how people use and experience space, how to put materials together, how to understand logic of composition, how to guarantee quality of construction, as well as how to dialogue with context of the site. On the basis of solving these fundamental questions, rising from technique level to aesthetic form level is the poetics of building. Furthermore, it demonstrate cultural spirit of the times and bears the speculation of life at the mental level. Ideas and thoughts are solidified within architecture. Architecture is used to reveal the hidden structural philosophy and aesthetic theory, which is also the eternal subject that architecture has been in quest of. There is an old saying in China, "what is within form is called physics, while what is above form is named as metaphysics." Scientific and rational ways tend to solve the first question in architecture, but the perception of ideology tries to tackle the latter question. The relation between "physics" and "metaphysics" in the field of architecture are complementary and inseparable.

Architectural Expression

表形与表意

建筑以形态造型为主要特征,形式语言是实现构思、传达讯息、表达思想的载体。形式的表达涉及形式的风格特征,风格的差异不会影响美感的表达。任何风格的表达都有层次的高低,文学作品中面对同一主题,可有不同的叙事方法和不同的风格表达,文笔或辞藻绚丽,或风骨苍老。同理,文化的多元带来建筑风格多样性,建筑的形式表达也形态各异,或古典严谨理性,或抽象现代感性。艺无古今,品有高下。无论何种表现方式和风格特征,见高下的是作品的思想内涵以及对格调极致的追求。

建筑的表达不同于文学语言的无穷表现力,文字语言的表达中语符的书写形式为能指(signifier),词语的表达对象、语符的意义或概念上的对应物为所指(signified)。"能指和所指"两部分构成语言的完整表达。两者之间的联系是相对应的,当建筑的"能指与所指"产生重合时,建筑之意义即建筑本身,建筑只能表达自己。只有表达的意义(meaning)即"能指"与"所指"建立起一一对应的指向关系,表意的具体形式通过视觉及自身体验为受众所感知并产生共鸣,才能实现由"立意表达"到"形态表现"完整的信息传递过程。

序 言 ……… 另壹种 表达

建筑的表意有别其他艺术的表达。建筑的造型手段受到功能使用、建造的技术以及经济因素的限制，建筑不能像雕塑、绘画等纯造型艺术那样具体地描绘特殊对象。但表达可以借助其他的艺术手段，来弥补建筑自身表现能力和手段的局限。

言授于意、意授于思、意在笔先是任何艺术表达的基本准则。立意通过形式语言来表达实现，是形式表达的出发点，形式是立意的最终实现结果和表达载体。建筑的立意来自不断变化的世界所带来的切身感受，以及对事物个性的理解和价值判断。同时建筑的立意应有准确的建筑表达，俱以意为主，意犹帅也，无帅之兵，谓之乌合。建筑的表达应该是娴熟畅达、由此达彼、言意相随的完整过程。形式的表达是建造的结果而不是建造的目标。建筑也不该成为平庸的形式机器，符合建筑自身逻辑规律的形式表达与以形式为目标是不同的概念。问题应该从属于结果，若把建筑形态附会成一个主题，针对形式的结果来寻找对应的问题，恰恰是本末倒置的做法。建筑的表形不能超出建筑自身的范畴与使用功能相脱节。建筑形式是立意和思想表达的目标和结果。

建筑的科学与理性因素构成了建筑学"形而下"层面的问题。思想意识形态层面的感性内容属于建筑更高层面的"形而上"问题。"形而下谓之器，形而上谓之道"。建筑的"道"和"器"的关系应该是相辅相成、互为表里的。

Presentation of Form and Meaning

Architecture features with form building. Formal language is the embodiment of idea realization, message delivery, and mental expression. Formal expression is related to formal style and characteristics; style difference doesn't affect aesthetic expression. Style expressions have diverse levels. The same literature subject could be presented by different narrative methods and various style expressions – either rhetorical or sophisticated. In a similar way, the diversity of culture brings forth the variety of architectural style. Formal expressions of architecture vary – they could be classical, precise and rational; or abstract, modern and sentimental. There are no ancient or modern arts, but only good or bad works. That is to say, whatever the expression or style is, the idea and connotation of the work and pursuit of class matters the most.

序言 ··—·. 另壹种 **表达**

Architectural expression is distinct from literature language with infinite expressiveness. In the literature language expression, the written form of a sign takes is signifier; the object the sign refers to, the meaning of the sign, or the corresponding concept it represents is the signified. The signifier and the signified, constitute the complete expression of language. In relation to one another, when the signifier and the signified is overlapped in architecture, the meaning of architecture is the same of the form of architecture. Architecture could only express its own. Only if the meaning of expression, i.e. the signifier and the signified, set up one-to-one correspondence, the concrete form of presentation of meaning is visually and bodily perceived by the audiences who are really turn in to them. In this way, the whole transmission process of information from "ideological expression" to "formal expression" could be fulfilled.

Architectural presentation of meaning differs from other kinds of art. Formative methods of architecture are restricted by function, building technology and economic factors. Therefore, architecture isn't able to specifically describe the particular object in the way that sculpture, painting and other pure formative arts do. Nevertheless, architectural expression could use other artistic means to make up for the limitation of its own expressive capability and means.

Language is determined by meaning; meaning depends on idea; thinking should be mandatory before speaking; those are the essential standards of any artistic expression. Presented by formal language, conception is the start point of formal expression. Form, is the final result of conception and the medium of expression. Architectural concept originates from the experience of the changing world and the comprehension of individuality and value judgment. Meanwhile, architectural concept should be expressed precisely. Meaning conquers like the marshal; corps without the marshal is just rabble. Architectural expression should be the complete process of skillfulness, fluency, logicality and coherence. Formal expression is the result of building, instead of aim. Architecture should not be a mediocre formal machine; formal expressions complied with architectural logic and forms as aim are different concepts. Question is subject to result. It would be like putting the cart before the horse if architectural form is regarded as a subject and seeks corresponding question for formal result. Architectural presentation of form shouldn't exceed its own boundaries and detach it from practical functions. Architectural form becomes the aim and result of conception and expression.

另一种表达

西尼·芬克斯坦在《音乐表达思想》一书中阐释了如何用音乐语言表达思想，他认为用音乐语言表达有时比文字更明确。门德尔松也常把他的无词歌曲寄给家人，作曲家认为音符不仅可以代替文字表达情感、心境，还能表达自己的思想和审美意象。

音乐如此，雕塑、绘画等其他造型艺术也都有自己的表达形式，并以职业为特征。不同的造型艺术载体成为对话的媒介，使思想的表达成为艺术的表现内容，唤起内心的情感意象来表达个人的主观感受。

建筑不仅是一个工程概念，还有人文属性的文学特质。作为有形的物质载体的建筑，在解决技术和功能使用问题的前提下，也能审美地表达个性的思想和主张，成为建筑师职业性的一种表达。本书选取表达不同主题的五项建筑设计编撰成集，用建筑的方式表达不同命题、不同主题的理解认识和个性解读。

潘家峪惨案纪念馆：在完成一个完整历史叙事、再现历史情景的同时，通过建筑表达了对历史事件的现实思考，以及对现实的积极启示意义。毋忘国耻、警钟长鸣、"前事不忘、后事之师"成为纪念馆要表现的主题。

井陉矿万人坑纪念馆：营造了一个展示史料、凭吊纪念的场所，建筑以荒芜的人工场景、浮雕和雕塑的表现，作为悲愤情绪的回应，以"生命"为主题刻画表现不屈的灵魂和求生的情景片段，表达对逝者的无限同情，对法西斯草菅生命暴行的痛恨，表达正义的情感。两个纪念建筑表达两个不同历史事件的不同纪念主题。手法上都借助雕塑语言及符号等其他表现方式，来弥补建筑自身表现能力和手段的局限。

建在采煤沉陷区的**唐山南湖展览馆**，展出的流线组织借用拓扑学中单侧面循环的"麦比乌斯圈"来诠释无限循环理念；建筑的构件、可回收的建造材料都表达建筑、改造的区域以及城市应该遵循的循环理念。

城市公园中山脚下的办公楼用"潜望镜"的意象作为建筑的形态特征，体现建筑的使用性质。最大限度地利用场地的环境特征，将建筑与所在环境融为一体，使建筑真正成为能"看得见风景的房间"。

当下急剧推进的城市化进程中，如何延续城市文脉、保存城市记忆，使建筑和城市更具有文化的内涵和品位？由旧粮库改造而成的**唐山城市展览馆**成为一个例证。通过城市空间重构与功能重组，展览馆重新营造了城市公共开放空间。通过旧建筑改造项目的"寓旧于新"理念表达的不仅是建筑策略，更应该是城市的策略。

建筑虽能直接或间接地表达一些思想，但建筑自由挥洒想象的空间是有局限的。真正能影响城市健康发展的，应该是城市规划的积极作用。规划对城市的作用要远大于建筑；现实社会中城市规划的制定受到政治、经济、技术、文化等多种复杂因素的制约和影响。核心价值观和思想理念影响规划策略的制定，规划的宏观综合性及复杂性特征导致实施过程的不确定性，城市规划总会成为体现权力的物质道具、获取商业利益的工具。城市规划专业所追求的空间理想、社会理想以及主流价值观的表达难以实现。

进入公共管理领域的专业人员都有追求专业理想的基本信念；城市规划的政策属性关乎决策的正确性与有效性。当技术立场与社会现实、政治、经济利益等诸多因素发生冲突时，往往会是专业立场失守，技术原则甚至变得可有可无。专业立场所关注的是技术理性和工具理性，而行政体系注重的是执行力、效率和全局性的整体观。城市规划的理想愿景表达面临诸多冲突和纠结，技术理想会逐步向实用主义的角色转换。

专业技术角色成为行政体系中的执行者，变成行政机器中的一个齿轮。城市规划的实践环节受多种复杂因素的制约，难以通过规划的方式表达自己个性的主张和立场，也许以更直接、可控的具体建筑表达更为顺畅。把城市层面的认知思考的观点主张也通过建筑的方式表达出来，建筑成为另一种表达。

A DIFFERENT EXPRESSION

Sidney Finkelstein elucidated how to express ideas using musical language in his book How Music Expresses Ideas. He said sometimes musical language is much clearer and more effective than words expression. Mendelssohn used to mail his music scores without lyrics to his family, too. The composer believed the note can replace words to express not only feeling and mood, but also idea and aesthetic intention.

Like music, sculpture, painting and other form arts have their own expression form, featured with professions. The carriers of various form arts are the media of conversation, turning the expression of idea into the content of artistic representation, and expressing the author's subjective conception by evoking inside feelings.

Architecture is not only an engineering concept, but also literally characterized with humanistic attribute. Under the condition of solving technical and functional questions, architecture, as the visible material embodiment, is able to aesthetically express personal idea and proposition. Thus, it becomes the professional expression of architects. This book selects five architectural design projects with diverse subjects, using architectural means to express various themes, perceptions and comprehensions.

Panjiayu Massacre Memorial Hall —By completely reproducing historical scenes, the architectural design presents the realistic thinking of the historical event and positive enlightenment to the present world. The main themes of the Memorial Hall are: to remember the nation's mortification, to keep alert, and to understand today is yesterday's pupil.

Jingxing Coal Mine Mass Grave Memorial Hall—It builds a memorial place to exhibit historic documents and to express condolence. In response to the grief and indignation, the building uses expression of artificial desert scene, relief carving and sculpture, describes the unyielding souls and scenes striking for survival themed with life. To show the never-ending sympathy, hatred to fascist killing the innocents, and show the rightful emotions, although the two memorial buildings have different subjects for two different historical events, they both adopt sculptural language and symbols, etc. to compensate for the limitations of architectural expressions and means.

Tangshan Nanhu Exhibition Hall —Built in the mining subsidence areas; the building's exhibition circulation adopting the concept of the one-sided and non-orientable surface, named Möbius strip in topology, to explain the infinite loop. Architectural components and recyclable building materials indicate the recycling idea that should be followed by building, renovated area and city.

The office building at the foot of the hill in the urban park adopts the periscope form to symbolize the building's functional properties. Maximize the site's surrounding features, integrate the architecture with the environment, and therefore make the building **A Room with a View**.

In contemporary radical urbanization process, how to extend the urban context and preserve the urban history so that a more distinctive cultural ethos can be embodied in architecture and city? **Tangshan Urban Planning Museum** renovated from an old granary is a good example. Through urban space reconstruction and function recombination, the exhibition hall represents urban open space. What have been expressed by the renovation project —retain the original within the new building —are not only architectural tactics, but more like urban strategies.

Although architecture is able to present some ideas directly or indirectly, the free architectural imagination space is still limited. It is the positive effect of urban planning that really influences urban healthy development. Planning contributes much more than architecture to the city. Urban planning legislation under present situation is restricted and affected by politics, economics, technologies, culture and many other complicated factors. The mainstream value and ideology influence the planning strategies formulation; the macroscopic comprehensiveness and complexity lead to the uncertainty of implementation process. Urban planning is definitely going to be the substantial property of power and tool for making commercial profit. The expression of space concept, society idea and mainstream value are very difficult for urban planning professionals to achieve.

Professionals who have entered into public management field have the basic belief of pursuing the specialized dream. The properties of urban planning policies concern the correctness and validity of the decision. When technological standpoint conflicts with social realities, politics, economic benefits, etc., it usually loses the game and technological principles sometimes get dispensable. Professional standpoint focuses on the rationality of technologies and tools, while administrative system emphasizes on implementation capacities, efficiency and integrity of overall importance. The idealistic scene expression for the urban planning faces many conflicts and entanglements. The technological ideal will gradually approach a way of pragmatism.

When professionals become the executers of the administrative system, they are in fact the gears of the administrative machine. In practice, the procedure of urban planning is restricted by various complicated factors, so it's rather difficult to present one's own proposition and standpoint by means of planning, but it may be relatively easier to adopt the more concrete ways as architectural expression. For this reason, the views and propositions of urban-level thinking are conveyed by architectural means. In this sense, architecture becomes a different expression.

|纪念主题的建筑表达| 另壹种 |表达|

1

Panjiayu Massacre Memorial Hall—architectural expression

潘家峪惨案纪念馆
纪念主题的**建筑表达**

潘家峪位于河北省丰润县，四面环山、山青水秀、民风淳朴。

潘家峪惨案是抗日时期发生在冀东地区的一个重要历史事件。侵华日军将全村的男女老少都集中到地主大院"潘家大院"并将大院团团围住，四周架设机枪、柴草进行血腥的灭绝性的屠杀。为纪念在惨案中死难的同胞，河北省政府决定兴建"潘家峪惨案纪念馆"，作为河北省爱国主义教育基地。旨在以历史的资料遗物和见证进行生动的爱国主义教育，教育国人及子孙后代牢记日军侵华史中无辜百姓被屠杀、同胞受屈辱的一幕。

纪念馆的选址在惨案发生地"潘家大院"一河之隔，设计把一直作为历史的遗迹文物加以完整地保留的"潘家大院"组织在纪念馆的展示流线中，平面布局以"院落（COURT）"为基本原型，用石头墙围合的院落式布局,形成沉重、封闭的院落空间。院落环线的空间序列组织展线并成为建筑构成的基本骨架，把遗迹作为整体展示的重要的组成部分，呈现现场的真实感。

用粗犷的毛石墙形成一种夸大的纪念尺度，表达历史的沉重感，将纪念馆的主入口设在院落的最窄处，刻意形成一种空间的紧张感，成为封闭的院落环境中绝处求生的一种空间暗示。庭院的景观处理更刻意以情景表达历史片段。院内石板与卵石铺成凶险而突兀的「爆炸形」，由具体意象的枯树、抽象夸张图形的铺地来描绘场地一种寸草不长的荒凉感，隐喻大屠杀后「白骨露于野，千里无鸡鸣」的场景，表现日寇屠杀无辜、惨无人道的「三光政策」。塑造一种沉闷、压抑的氛围，强化人们的心境和感受：大地遭洗劫，空旷、寂静、萧瑟、没有生命、没有哭声与叹息。恍若无数魂灵在沉默中控诉，天地为之大悲而无言。

纪念主题的建筑表达 | 另壹种 表达

纪念馆的内部展览用流畅而简洁的空间来布置流线，空间尽量将建筑的本体空间维持平实而沉稳且具有弹性发挥的中性角色，不明显强调花俏的语言与空间样式逾越展示内容本身。将诉说的空间留给陈展的历史主题，用图片、文字材料等媒介组合表现多重的叙事和情境。

纪念馆用建筑的语言表现这一历史事件并表达正义的情感。展览用大部分展线渲染凄惨、悲壮的场面，把历史浓缩、凝固在空间里，让现实的人们牢记发生的历史事件，并展现人类历史上令人发指、令人心悸的一幕。刻意营造令人压抑的氛围，从而唤起同胞对这些悲难的历史记忆。建筑成为一个记载悲痛与缅怀历史的纪念场所。

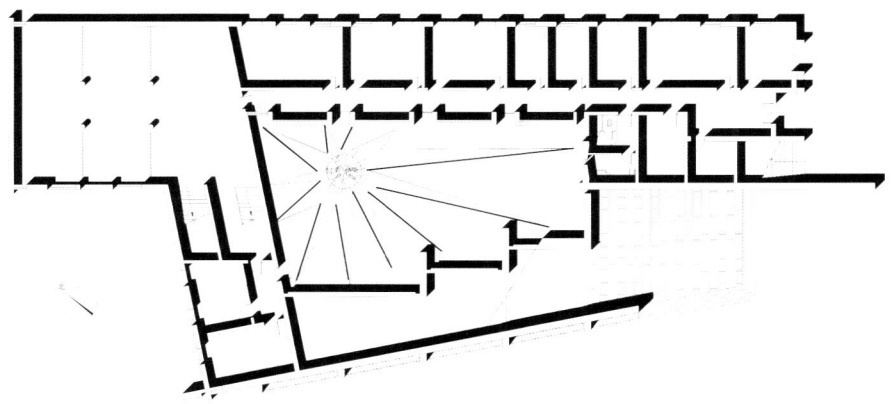

纪念主题的建筑表达 ——另壹种表达

对于纪念一个悲惨事件主题的建筑空间而言,"让过去成为过去"的概念是消极的,所以展线的展示再现悲惨事件的同时,并没有一味展示宣泄这种悲壮的气氛,而是控制展览的节奏变换。纪念展示历史事件的同时,更要表达纪念主题的积极意义。

纪念馆拾阶而上迎面的实墙上构图需要一个景窗,我们将窗上图案变形处理成一个镂空的"耻"字,一、二层展厅转换的室内楼梯间也会再现"耻"字的投影,牢记历史、不忘国耻。只有"知耻"才能"后勇"。日本侵华战争是我们民族的"国耻",带给我们的深刻教训是积极地提醒这样的耻辱历史不再重现、永不发生。

展览最后结束时步出展厅,迎面为两个刀型组合而成的垂直钟塔,随风摇曳的铜钟,不时被敲响并传遍整个村落,"警钟长鸣"的寓意不言而喻。钟声警示人们不要忘记发生在这里的悲惨一幕,日本侵华战争带给我们民族深重的灾难,成为我们民族的耻辱;警示人们不要忘记我们民族曾经受屈辱的历史。后人有责任对历史保有一种清晰的记忆,同时对现实保持一种敏锐,永远记住落后就要挨打的教训。"前事不忘、后事之师"引发人们对现实及未来的思考。这个纪念馆最终要表达积极而现实的意义,这也正是纪念馆刻意要表现的主题。

纪念主题的建筑表达 ········ 另壹种 表达

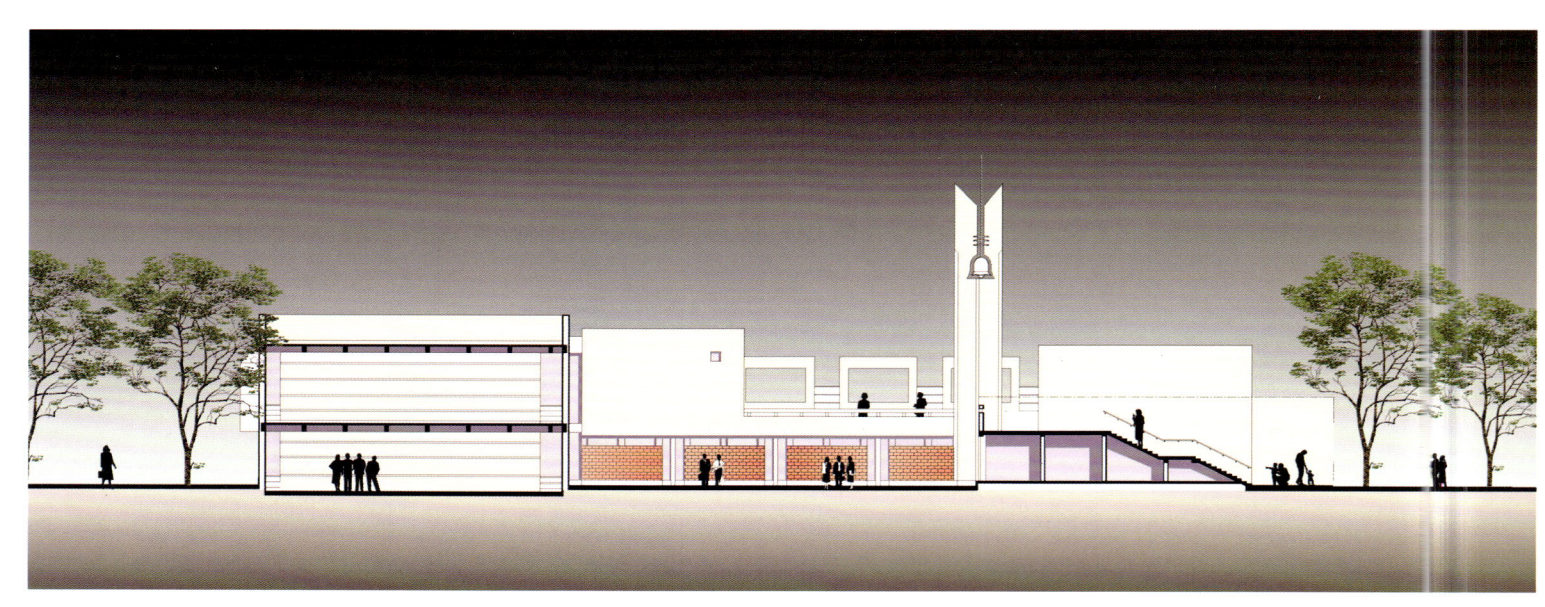

Panjiayu is located in Fengrun County, Hebei Province. It's surrounded by numerous mountains and has picturesque scenery. The Panjiayu people are stocky and honest.

Panjiayu Massacre is a big historical event taking place in east Hebei area during anti-Japanese war. The Imperial Japanese army herded all the people in the village to "The Pan Compound" and blockaded that compound. They set up machine guns and hay, and carried out bloody carnage. To commemorate those who died in tragedy, Hebei Province government decided to construct "Panjiayu Massacre Memorial Hall", as the provincial patriotism education demonstration base. It aims to display historical relics and witness to teach patriotism, teach the people and future generations to remember the scene when innocent people were killed and humiliated in the history of Japanese invasion of China.

The Memorial Hall is located at the other side of the river near "The Pan Compound" where the massacre took place. The design keeps the well-preserved "The Pan Compound" in the Memorial Hall's exhibition circulation. The plan layout uses "COURT" as the prototype; the courtyard surrounded by stone walls forms a heavy and enclosed space. The spatial sequence circling the courtyard organizes the circulation and becomes the elementary skeleton of the architecture. The relics are displayed as the major parts of the whole exhibition, demonstrating the field realness.

Rough rubble walls are used to form an exaggerate memorial scale to present the heaviness of history. The main entrance of the Memorial Hall, set at the narrowest of the courtyard deliberately producing a tense atmosphere, is the space metaphor of desperate play in the enclosed courtyard. The landscape design of the yard emphasizes on situational presentation of the historical period. The flagstone and pebble shapes dangerous and abrupt "explosive view" ; the dead tree and other astonishing abstract paving patterns are used to describe the infertile desolation, to express the scene after the massacre. For example, white bones are exposed all over the field; no cock crows could be heard within a thousand miles. This scene was resulted from Japanese army's inhumane "Three Alls Policy" (Kill all, burn all, and loot all) with an aim of cruelly killing innocent. Such landscape design seeks to create a dull and depressing atmosphere, and generate the feelings to facilitate memories. The land was robbed; open, silent, withered, lifeless, no crying or sighing. It's just like numerous ghosts are accusing in the silence; heaven and earth are speechless, filled with sorrow.

纪念主题的建筑表达 ········· 另壹种 | 表达

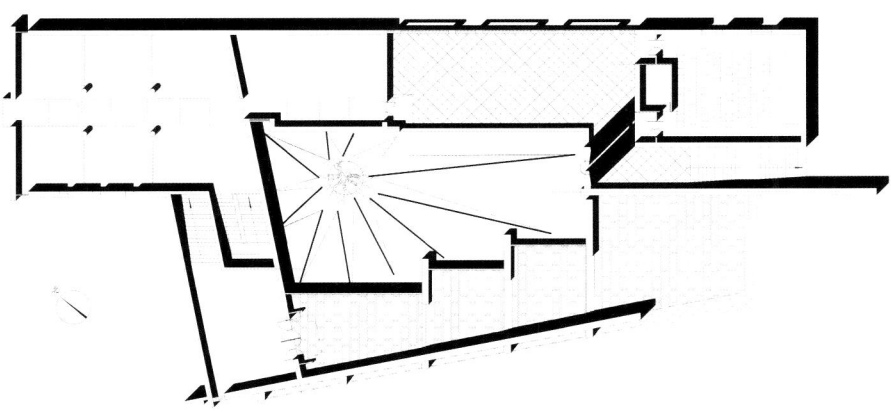

纪念主题的建筑表达

另壹种 **表达**

The interior exhibition of the Memorial Hall adopts a simple but smooth circulation to organize spaces. The space makes an effort to keep the building simple and calm, playing a flexible neutral role. Flaring language and space pattern are not emphasized to avoid going beyond the content, leaving space for the historical theme of exhibition. Photographs and text materials and other medium are combined together to present a multiple narrative situation.

The Memorial Hall uses architectural language to present this historical event and express rightful emotion. Most of the exhibition circulation is designed to render the miserable and tragic scene, condense and solidify the history into space, enable people nowadays to remember the historical event. The shocking and horrifying scenes in the human history are represented to produce depressive atmosphere, thus to remind the compatriots of those dreadful historical memories. Architecture becomes a memorial place for representing sorrow and commemorating history.

In regard to the architectural space with the tragedy memorial subject, the concept of "letting the past pass" is negative. So, when the circulation reproduces the massacre, it doesn't only focus on giving utterance to tragic feelings, but aims to control the rhythm alteration of the exhibition. While commemorating and exhibiting historical event, the architecture design makes efforts to express the positive significance of commemorating subject.

纪念主题的建筑表达 ······ 另壹种 表达

The wall facing the Memorial Hall's staircase needs a view window for contrapposto, so we designed the pattern as a hollowed-out Chinese character "耻" (mortification), and the interior staircase connecting the first floor and second floor will be projected the "耻" (mortification). Remember history and never forget national mortification. There is a Chinese saying: knowing the things of shame is to be near fortitude. Japan's aggression against China is our nation's mortification, which teaches us a lesson we shall never let this happen again.

Stepping out of the exhibition hall, you'll see a vertical bell tower combined of two knife-shape components right in front. The bronze bell swinging in the wind is being struck now and then, going around the whole village, underlying the metaphor of "the alarm sounds continuously". The bell sound warns people not to forget what has happened here. Japan's aggression against China brought our nation serious disaster and deep shame, so the design has to remind people not to forget this period of humiliated history. Offspring is responsible to keep clear memory of the history, and meanwhile be keen to the reality. Remember that a nation is vulnerable to attack if lagging behind. "Today is yesterday's pupil" initiates people's thinking of reality and future, which is the Memorial Hall's ultimate positive and realistic meaning, and also the subject Memorial Hall trying to present.

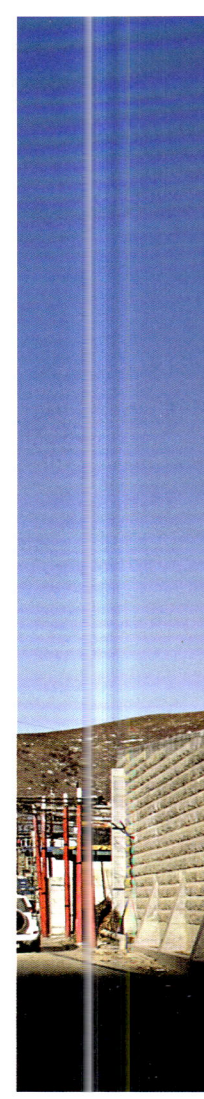

2000年全国优秀设计评选专家评语

根据建设部《关于开展2000年度优秀勘察设计评选活动的通知》，中国勘察设计协会受建设部的委托，2000年9月14日在北京组织29位各方面专家，进行了部级优秀建筑设计的评选工作。这次评选的建筑均为1999年7月以前竣工的工程，主要是90年代后期的，反映出我国20世纪末的建筑设计水平。比起前几届参评的作品，本次作品确实反映出我们的建筑设计水平不断在提高，日益成熟，建筑创作的道路也日益宽广。新人、新作品大量涌现，令人欣喜振奋。

改革开放以来，社会为建筑创作提供了巨大的物质基础，为建筑设计提供了更大的动力和更多的机遇；但同时也带来了商品化和强势文化的强烈冲击，我们的建筑设计作品也都或多或少地反映出这种影响，到目前这种影响依然存在。但经过这些年的评论、探讨、实践，建筑师们正在积极地对待各种矛盾，面对现实，努力创造出新时代中国的建筑作品，有不少作品取得了很好的综合创意效果。这次参选作品中有不少堪称"精品"；从设计的布局，到空间、造型、环境等都是精心推敲、反复研究的。如果没有精品意识，只是随波逐流或照章办理，是做不出优秀作品的。例如潘家峪惨案纪念馆虽然是个1240平方米的小型纪念馆，设计者没有遵循那种"庄重"、"华贵"的套路，而是将潘家大院这一真实的历史遗迹与纪念馆组织在一起，用建筑空间组织了参观路线，引导视线强调主题。用粗犷而经济的地方石料作饰面，很好地体现了纪念馆的个性。尤其值得一提的是，其造价不过每平方米1611元。

——专家组评委　北京建筑设计研究院总建筑师　　魏大中

每次参加部优秀建筑设计评选，不但见到许多设计精品、了解我国整体设计水平和发展动向，而且又是一次学习的好机会。我参加过许多届优秀设计的评选活动，尤以这一届的成果最令人鼓舞，出现不少亮点，主要表现在两个方面：一、作品的创新力度大，有时代气息。一些优秀作品给人以耳目一新的感受，其中，尤以博物馆和文化建筑最为突出。二、中青年建筑师茁壮成长，许多有创意的优秀作品，如潘家峪惨案纪念馆、天津师范大学艺术体育楼、外语教学与研究出版社办公楼等一批优秀作品，都出自"文革"以后培养出的年轻建筑师之手，他们逐渐成为我国建筑界的主流。这是我国建筑发展的希望所在。建国以来，尽管我们的设计能力有了很大提高，也出现了一批好作品，但总体设计水平与国外先进国家相比，在城市与环境协调、建筑空间组织、新材料和新技术的运用以及建筑细部处理等方面还有不小差距。要创作有中国特色的现代建筑，我们的建筑师在理论和素质上，还需要下很大的功夫。要重视建筑的整体观和可持续发展观。在建筑创作上要强调地域性、文化性、时代性，只要沿着这个方向不断探索，就一定能创作出更多有创意的优秀作品。

——专家组评委　中国工程院院士　华南理工大学建筑设计研究院总建筑师　何镜堂

潘家峪惨案纪念馆

2000年荣获第九届国家优秀设计银质奖、2000年建设部全国优秀工程设计评选一等奖。2009年获新中国60年来各时期我国建筑创作优秀代表性作品的最高奖项"中国建筑学会建筑创作大奖"。

| 历史事件的主题表达 | 另壹种 表达

2

Jingxing Coal Mine Mass Grave Memorial Hall—expression of historical event subject

井陉矿万人坑纪念馆
历史事件的主题表达

20世纪初德、日法西斯侵华时期，在河北石家庄井陉煤矿持续40余年开采煤矿，大肆掠夺资源，欺诈剥削中国劳工。开采井陉煤矿过程中陆续有大致46000名死难的矿工被埋在被称为"南大沟"的地方，许多丧失劳动能力还尚有生命的矿工被活埋于此。在现场遗址处挖掘出的尸骨惨状怵目惊心，令人不寒而栗。2001年河北省委宣传部将井陉"万人坑"的遗址辟为纪念性的主题公园，建设纪念馆作为省级爱国主义教育基地，记载历史、警示后人。

纪念建筑有别于他类建筑，建筑的精神层面要大于功能使用层面上的要求，纪念建筑的主题表达如下：首先应是立意高远、主题鲜明；其次是如何运用恰当而有效的方法和途径来表现主题；再有就是传递与表达的信息可以被直接感知，从而引发观者情感上的共鸣。

采取何种途径和方法，恰当有效地表现纪念建筑精神层面的内容至关重要。建筑不可能像其他造型艺术那样可以具体地描绘特殊对象，只能借助于其他的造型艺术手段来弥补建筑本身表现不足的方法和途径。建筑的表现是抽象并有其局限性的，建筑本身的艺术表达应遵循建筑的逻辑理性与技术理性原则。建筑也不能与其使用功能相脱节，超越自身的范畴为表现而表现。纪念建筑无论采取具象或抽象的表达方式，所传递与表达的信息应该是可感知的。纪念主题表达应与情感互达、准确、易懂，并有艺术的感召力。

如何提炼纪念主题来纪念这一悲惨的历史事件？运用鲜明特色的建筑语言来营造情感空间，进而表达富有感染力的纪念主题，纪念非正常死难的劳工，「生命」就成为纪念馆所要刻意表现的主题。

将以「生命」为主题的雕塑、浮雕等视觉形象有机地结合组织在陈展流线与建筑的空间序列中，展示纪念的「生命」主题。室内展陈与室外场景和雕塑通过展线交替地组织在一起营造气氛。折返空间序列以延长有限的展线，通过展线来控制空间的转换与空间节奏。利用原有的地貌特征表现室外的场景，拾阶而上的平台一眼望去，一片冷漠的虚空，寸草不生。由黑色花岗岩分隔成层层叠叠与头骨相当的灰白卵石，对比强烈。黑色是煤的隐喻，数不尽的卵石犹如一个个鲜活的生命，财富与生命的关系不言而喻。漫山遍野的尸骨犹如山一般，成为凄惨情景的例证，且铁证如山！

历史事件的主题表达 ········· 另壹种 **表达**

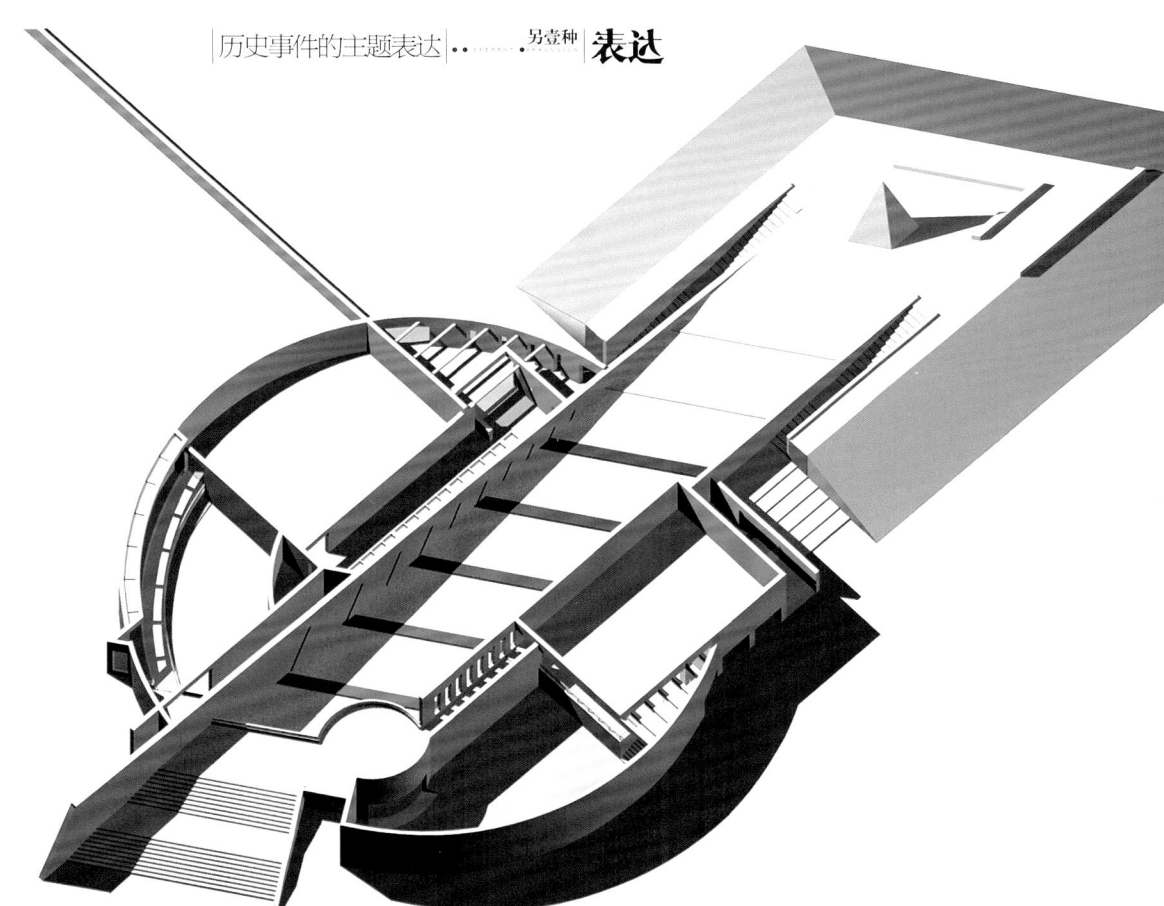

历史事件的主题表达 · 另壹种 表达

历史事件的主题表达 ……另壹种 表达

求生是人的本能,可以想象,生命尚息、被活埋的矿工或处在生命的最后的挣扎会体现在手的动作上。我们与雕塑家几易其稿,将无数挣扎残缺的「手」构成一组锻铜浮雕,放在建筑最显要的位置上,并处理成镂空状,夜景照明光由里及外,逆光下的一只只挣扎向上的「手」,表达出令人心悸的一幕。与此相呼应,尺度巨大并倾斜的「手的负形」成为纪念馆的至高点,也是空间序列的底景。将手的形式与生命联系起来,暗示掩埋在此的灵魂是空的,向上无助的手形、累累的「白骨」与逝去不屈的冤魂,一虚一实。量的叠加与虚空的处理使这一纪念主题得到强化。46000个不屈的生命、不散的冤魂,仍在挣扎和抗争。中国的传统观念会把人间分为「阴—阳」两世界。纪念主题的立意与表达也正在于此,也相信逝者在天有灵!

During the Germany and Japan's aggression against China back in the early 20th century, Shijiazhuang Jingxing Coal Mine in Hebei Province had been developing coal mines for over 40 years. The fascists over exploited resources on large scale; cheated and exploited Chinese labors. In the process of developing Jingxing Coal Mine, there were about 46,000 dead miners buried at "Nandagou", and even many incapacitated miners were buried alive. The skeletons excavated at the site are too scary and creepy to look at. In 2001, Hebei Propaganda Ministry built a memorial theme park on the site of Jingxing mass grave, and built a Memorial Hall as the provincial patriotism education demonstration base to record history and warn descendents.

Memorial buildings are different from other types of buildings. The demand on spiritual level exceeds the functional level. What the memorial buildings express are following subjects: first of all, it should be a profound concept and have a distinct theme; secondly, how to use the appropriate and effective method and means to express the theme; thirdly, ensure the conveyed and expressed information to be perceptive directly, therefore initiates the audience's resonance feeling.

历史事件的主题表达:另壹种表达

It's vital to decide what kind of method we use for accurately and effectively expressing the emotional and psychological contents of memorial buildings. Architecture is unlike other visual arts which could describe particular objects specifically, but only use some methods of the other visual arts to make up its own deficiency in expression. Architectural expression is abstract and limited; architectural art expression shall obey logic and technical principles in a rational way. Also, architecture shall not detach from its functions, shall not go beyond its own category to express.

No matter the expression is concrete or abstract, the information delivered and expressed by memorial buildings can be perceived and conceived. Memorial themes expression should be easily communicated, precisely conceived, and definitely understood, and so that they have evocative imaginary.

How to refine the memorial theme to remember this historical tragedy, and how to use featured architectural language to make emotional place, thus to present appealing memorial themes, to commemorate labors who died a violent death? "Life", therefore becomes the deliberately presented theme of the memorial hall. Sculptures and relief carving and other visual images with "life" memorial theme are organically combined in the exhibition circulation and architectural spatial sequence. Interior exhibition, exterior scenes and sculpture are weaved together by the circulation to create atmosphere. The zigzag spatial sequence extends the limited circulation, which controls space shift and rhythm.

Original geomorphology is used to present exterior scenes. Viewed from the stepped platform, it's cold and distant hollowness, not even a blade of grass grows. Layer upon layer of dark granite and skull-like white pebbles produce sharp contrast. Black is a metaphor for charcoal. Numerous pebbles are like fresh lives, and the relation between fortune and life is self-evident. Skeletons scattered over hill and dale, piled up like a mountain, explain the miserable scene. Ironclad proof is as real as the mountain!

Humans have an instinct for survival. Imagine that those dying buried-alive miners or some person struggling for survival, their last action must be to move their hands. We've discussed with the sculptor and revised the design several times, put many struggling incomplete annealed copper "hands" together to form a set of relief sculptures, and laid them at the most important spot of the building. The relief sculptures are carved hollow out. When the night light beats down outwards, the upward back lighting "hands" present a horrifying scene. Correspondingly, giant tilted negative image of "hands" become the commanding height of the memorial hall, and also the background of spatial sequence. Combining the hand form and life to imply souls buried here are void; while helpless upward hands form, piles of skeletons and unyielding wronged ghosts are existent. The memorial theme is strengthened by the accumulation of volumes and processing of void. 46,000 unyielding lives, haunting ghosts are still struggling and fighting. According to Chinese tradition, there are two different worlds- "yin" and "yang". The conception and expression of the memorial theme is right originated from it. We believe that the departed could sense it in heaven.

河北井陉矿万人坑纪念馆

2009年获中国建筑学会建筑创作大奖"入围奖"
新中国60年来各时期我国建筑创作优秀代表性作品

获奖证书

唐山市规划建筑设计研究院：

你单位设计的 井陉矿万人坑纪念馆 项目，荣获建筑创作大奖入围奖。特颁此证

中国建筑学会
二〇〇九年十二月

|环境的回应表达| |另壹种|表达|

3

Panjiayu Massacre Memorial Hall—architectural expression

看得见风景的房间
环境的**回应表达**

这又是一个小项目,要在城市公园的山脚下建办公楼。建筑的功能很简单,但场地特征挺有特点。基地一侧是由于采石而裸露的山体,基地的对面还有几棵长势不错的大树,这个题目要解决的关键是处理好建筑与山和树的关系,以求得建筑与场所之间的平衡。场地特质所激发的直觉,确立了新建筑应该把建筑做成山的一个整体,并最大可能地修复山体,再有就是完全保留现有树木,最大限度地利用现有树木并成为建筑内外的景观要素,由此衍生出建筑"看"与"被看"的关系。

建筑不仅是形象的表现,丰富的空间体验与有美感的建筑形体同样重要。建筑是人和他周围的世界发生关系的媒介,室内空间关系到使用者每时、每刻的感受,建筑师应该通过建筑表达自己对环境的理解,并把这种感受传递给使用者。

建筑的平面布局将建筑的主要办公空间成角度斜向布置，并处理成指向性强的筒体，每个筒体都面对要保留的树木，强调室内空间与基地特征的关联，表达出一种清晰的逻辑关系，使室内大部分空间都具有开阔而丰富的视野，最大限度地把室外树木引入室内。三个标准的办公单元筒体的两端为无框玻璃，犹如望远镜的两组镜片，标准单元内可看到室内公共空间，同时也可看到室外，具有方向感很强的两侧视野，公共走道处视线可穿过办公单元看到室外景色，其他办公空间也能看到不同视角的自然景观。

环境的回应表达——另一种 表达

一棵30年树龄的栾树位于会议室窗外,把这棵树作为室内重要的景观要素,将会议室悬挑2米最大限度地接近这棵树,在会议室内看到的恰巧是树木枝叶最优美的部分。试想在生机盎然的初春、枝叶茂盛的盛夏、色彩斑斓的深秋、大雪压枝的寒冬,人们都能透过玻璃真切地感受到这棵树的存在,通过这棵树感受到天气与四季的变换,室外的树木犹如一张四季变换风景画,影响室内的会议氛围,将人的活动与景观巧妙地融合在一起,来营造丰富而有趣的空间体验。

[环境的回应表达]……另一种 **表达**

环境的回应表达：另壹种**表达**

建筑形式是由内到外的自然显现，犹如从山里伸出来的几块巨石构成了建筑的主体。建筑形体处理简洁、明了，并符合山的尺度。"望远镜"筒体的外墙采用与山体色彩接近、竖贴的青石板表现材料的肌理与质感，强调主体形式的雕琢感。屋顶由植被完全覆盖，等五叶地槿爬满外砌毛石墙的墙面时，建筑就与山体浑然一体，建筑已成为山的延续。

在把握场地基本要素地形、地貌、景观的同时，更关注建筑构成的基本要素，把建筑做得更富情趣。简洁的材料纯化了室内空间，营造出一种不乏味的简约空间。材料强调色彩与质感的对比，强调材料之间的细部交接关系，来减少材料之间的转换过渡。有重点地在室内引入自然光，用光线来营造空间氛围，使光成为建筑的灵魂，提高了建筑的空间品质，同时赋予使用者以简洁、明彻的空间体验。

This is another small project to build an office building at the foot of the city park. The functions are simple, but the site is characteristic. One side of the site is the exposed mountain massif due to quarry; there are a few big healthy trees opposite the site. It's crucial to deal with the relation among architecture, mountains and trees, in pursuit to balance of the building and site. Instincts inspired by the site characteristics established that: the new building should be incorporated into the mountain, the massif should be restored as much as possible, and all the existing trees should be fully protected to be the landscape elements inside and outside the building. Thus the architectural relation of "view" and "to be viewed" can be developed.

环境的回应表达……另一种 表达

Architecture is not only the expression of image. Rich spatial experience is as important as physical beauty of the building, so architecture is the medium of people connecting their surrounding world, interior space relates to every moment of the user's experience. Architects should express comprehension of environment through their architecture projects, and pass on to users.

The major office space is arranged in a diagonal angle in the building layout plan, and designed as a directional cylinder. Every cylinder faces the retaining trees, emphasizing that interior space is associated with site features. The clear logical relation enables most of the interior space to have a broad and rich field of view, introducing the outside trees inside to the full extent. Both ends of three typical cylinder office units are frameless glass, similar to two sets of telescope lenses. Interior public space can be viewed from typical units, and so does exterior space. With a strong sense of direction on both sides of view, people at the public walkway could see the outside view through the office units. And, natural landscape could be viewed at other office space from different perspectives.

A 30-year-old golden rain tree outside the meeting room window is introduced as an important indoor landscape element. The meeting room cantilevers 2 meters to be as close to the tree as possible. What people can see from the room happens to be the most beautiful part of the tree branches and leaves. Imagine no matter exuberant spring, leafy summer, colorful autumn, or snowy winter, people can vividly feel the presence of the tree through the glass. Via this tree, we can feel the weather and seasons change. The outdoor tree is like a seasonally changing landscape painting, affecting the atmosphere of the meeting room, subtly integrating people's activities and the landscape together to create a rich and interesting spatial experience.

环境的回应表达 ········ 另壹种 表达

Like huge rocks sticking out from the mountains, the architectural form is the natural appearance from inside to outside, constitutes the main body of the building – concise and simple, and consistent with the scale of mountains. The external walls of cylinder "telescope" adopt vertically-pasted slab which color is close to the mountain body, to show the material's texture and emphasize the carving form of the main body. Roof is completely covered by vegetation. When Virginia creeper crawls over the rubble wall, the building and mountain look like an integrated mass; the building is already a continuation of the mountain.

While grasping the basic elements of site like topography, geomorphology and landscape, we should pay more attention to the basic elements of architectural composition at the same time, making the building more interesting. Simple materials are used to purify indoor spaces, and create an attractive minimalist space. The materials emphasized on contrast between color and texture, on the intersection of different materials details, in order to reduce the transition between materials. By deliberately introducing the sunlight, the architect uses light to create a spatial atmosphere, so that light becomes the soul of the architecture and therefore improve the quality of architectural space, at the same time endue the users with concise and bright spatial experience.

// 另壹种 | 表达

4

Tangshan Nanhu Exhibition Hall – architectural expression of cycle concept

唐山南湖展馆
循环理念的**建筑表达**

物质非一成不变，自然事物的发展和变化都是不断运动和相互作用的结果。相互对立、依存和转化是宇宙间万事万物生生灭灭的规律和原因。传统五行学说归纳世界是由金、木、水、火、土构成，循环往复、周而复始，并相生相克。其观点朴素且唯物。

以煤的形成为例：植物受光合作用成木、成林，深埋于地下的林木，成黑色可燃化石，变成煤炭，煤燃烧的同时又发出光和热，从而构成了物质世界的能量循环。

唐山的采煤沉陷区的更新改造也遵循这一理念。这里曾有良好的植被，经过自然造化，形成矿藏，后来，由于人类的开采，地表日渐荒芜，良田成为城市边缘荒芜的废弃地。斗转星移，如今的南湖，经过大规模的治理改造，湖光一色，鸟语花香，废弃地改造不仅凸显其巨大的生态价值，且已变成城市的中央公园，将来还会改变整个城市结构。南湖区域改造已成为城市可持续发展的范例。

在这一特定的场地建设展馆，成为我们用建筑设计表达循环理念、可持续发展的重要契机。展示改造的细节过程，记录下人与自然互动中的万千变化，展示对未来的美好愿景。

选址成为设计的开始，选择的场地不改变任何地貌特征，在两湖之间，前湖辽阔，后湖精致。长方体的展馆与道路旋转成一定的角度，面向湖长轴方向，并跨过道路，挑向湖面，原有的路依然保留。由于路的保留，建筑作了悬挑，与自然保持着积极且和谐的关系。作为对地形特征的回应，展馆的形态因此变得不同……建造的材料体现了循环理念，钢结构主体具备可拆改的条件，可循环使用。首先尽可能地减少建筑内外材料的种类。展厅内墙、地面与顶棚全部使用集合木板，弱化自身实体从而突出主题。材料尽量保持原有的本色和质地，自然演变的色彩比人工设色更具真实性，铁锈色在自然环境中真实协调，随着时间的推移，颜色会变得更暗，虽锈迹斑斑，却是更加真实、自然的变化。

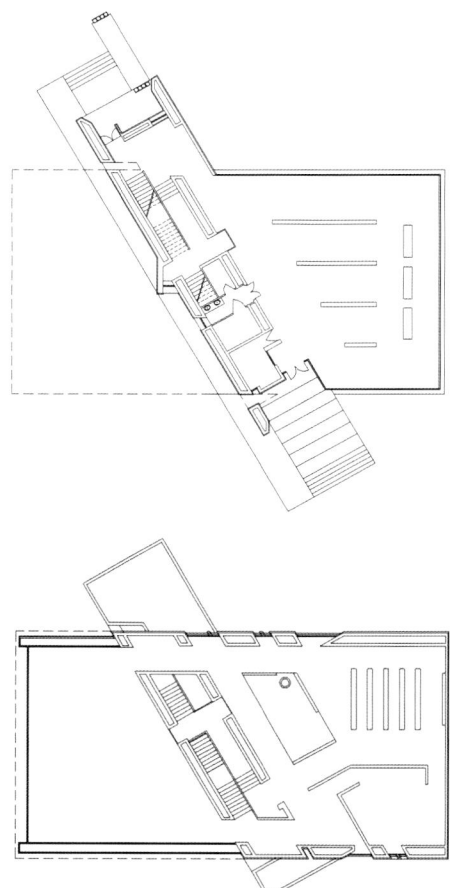

首层展馆的功能简单,陈展图片与文字的首层展厅三面透明玻璃接纳了周边全部的湖水,顶棚下垂的展板展陈历史的影像,旧影像与真实的自然景色在同一空间并置,形成「昔日倒影两重天」反差对比。拾梯而上,带着登高远望的心情,顺着光线进入由磨砂、镜面和透明玻璃构成的二层情景展厅,这里不需要展示任何物品,眼前看到的是水天一色的景色。自然是最好的展品,也是建设成就的奖杯外,这里唯一的展品。镜面玻璃反射出对面的真实风景,与虚幻的反射集合在一个空间内,亦真亦幻,每一天的日更月替,每一季的春去秋来,不断循环往复的风景成为最生动、最具灵性的展品。二层的演播厅,与情景展厅的明亮相比,这里的暗更有利于电子演示,窗在墙面上有节奏的切分,形成了附近和远处景观的框景,并且展现出光线的运转和流动。有的能采集一缕晨光,有的正对一片湖面。

展馆设计像一个高效的展示装置,展示的路线设计借鉴了「麦比乌斯圈」无限循环的概念,将拓扑学中最有趣的单侧面循环做法与楼梯分层组织,互不交叉又上下行的两股人流,空间上形成四个不同的标高,串连景色与展品,将内外空间流畅地组织起来。用建筑的参观流线直接地表达了无限循环的生态理念。

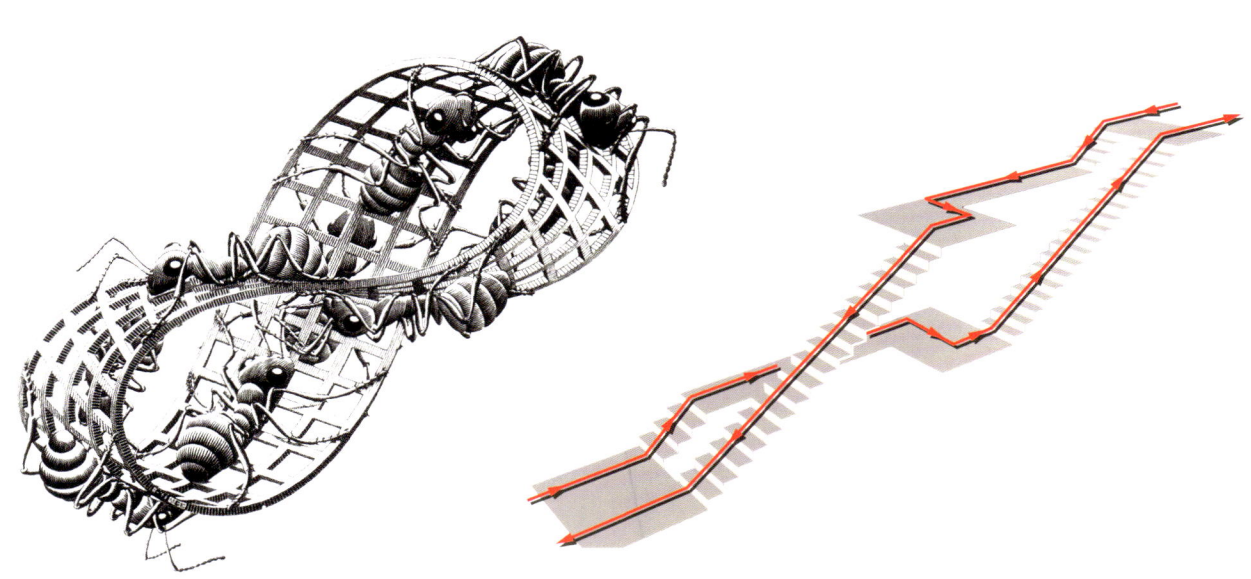

循环理念的建筑表达 ········ 另一种 **表达**

All things are changeable. The development and changes of natural substances are the result of constant motion and interaction. Contrariety, interdependence and transformation are the disciplines and causes of everything's life cycle in the universe. The theory of five elements summarizes that the world is constituted of metal, wood, water, fire and earth, which repeat themselves in endless cycles and counteract each other. It's a simple version of materialism.

Take coal's formation for example, plants turn into wood and forest by photo synthesis; trees buried underground grow into black flammable fossil, then into charcoal; charcoal gives out light and heat while burning; which constitute an energy cycle in the material world.

循环理念的建筑表达……另一种 表达

循环理念的建筑表达 ········ 另壹种 表达

南湖展馆
南湖 PARK 2006

Tangshan sinking land in coalmining areas' renovation also follows this philosophy. There used to be good vegetation, and then mineral resources are formed through natural transformations. Afterwards, due to human exploitation, the earth surface gets increasingly barren; fertile land becomes desert wasteland in the suburb. With the passage of time, today's Nanhu has stunning lake view and very good ecological environment through large-scale management and reform. The wasteland renovation not only highlights its tremendous ecological value through turning into an urban central park, but also will change the entire urban structure in the future. Nanhu area renovation has become an example of sustainable urban development.

Building the exhibition hall on this particular site is an important opportunity for expressing recycling concept and sustainable development in our architectural design. Make efforts to show details of the renovation process, record myriad changes within man's interaction with nature, express wonderful vision for the future.

Location as the start of the design, the site between two lakes does not change any topographic features. The front lake is vast; the back one is delicate. Rectangular hall is at a certain rotation angle of the road, facing the prolate axis of the lake and across the road. Since the original road is retained, the hall stretches toward the lake, cantilevered and keeping a positive and harmonious relation with nature. As a response to the terrain features, the hall's form is changed accordingly...

Building materials reflect the recycle concept. The main steel structure is recyclable, which can be deconstructed and renovated. First, we should minimize the types of building materials both inside and outside. The inner walls, floors and ceilings of the exhibition hall all use plywood, weakening their own identity to highlight the theme. They are constructed with the materials in authentic colors and textures; the natural evolution of colors is more realistic than artificial colors. The corroded steel plate of external walls better reflect the material properties. The rust color truly coordinates with the natural environment. With time passing by, the color will get darker. Although it's rusty, it's more real and reflects natural transformations.

The first floor of the exhibition hall has very simple function. Displaying photographs and text materials, the exhibition hall is fitted with transparent glasses on three sides, which take in all the surrounding lake view. The pendant ceiling panels showcasing the historical images; the old images and the real natural beauty are juxtaposed within the same space, forming a contrast comparison of "double reflection of the old days".

Stepping up the stairs, in a mood expecting to gaze far into the distance, following the light, visitors enter into the situational hall encircled with matte, glossy and transparent glass on the second floor. Except trophies representing construction achievements, there needn't to display any other items. What's before visitors' eyes is the view that the water and sky blend in one color. Nature is the best, and the only, exhibits. Specular glass reflects the real view of opposite, gathering with the visional reflection in the same space. It's in between reality and illusion. The view of everyday passing by and seasonal changing and recycling is the most vivid and spiritual exhibits.

Compared with the bright situational hall, the studio on the second floor is dark, which suits for electronic presentations. The windows leave rhythmic segmentation in the wall, forming a framed view of nearby and distant landscape, and show the light's movement and flow. Some can capture a ray of morning light, and some are right facing the lake.

Exhibition design is like an efficient display installation. Learning from the infinite recycling concept of "M bius strip", the exhibition circulation design adopts the most interesting one-side recycling practice of topology and uses staircases to part the people flow going up and down every floor. Therefore, it forms four different elevations connecting scenery and exhibits, smoothly organizing the interior and exterior spaces. The design uses a visiting circulation to directly express the infinite recycling ecological concept.

唐山南湖展馆
——南部采煤沉降区治理项目

南湖地区是唐山最大的采煤塌陷区,是开滦唐山矿采煤沉降区,曾因污水四溢、满目狼藉而被称为"龙须沟"。紧邻城市中心区, 2004年8月由这片采煤沉降区改造而成的南湖公园,获得联合国人居署(UN-HABITAT)可持续发展人居领域最高的奖项"联合国人居署迪拜最佳范例奖"桂冠。如今这里碧波荡漾,绿树成荫,已成为中国煤矿城市"采煤沉陷区治理"城市环境改造的经典范例,并成为唐山城市转型的象征。"联合国人居署迪拜国际范例奖"以其严格的评审标准,成为全球在可持续性发展方面做出突出贡献的"桂冠"奖项。

| 城市认知的建筑表达 | 另壹种 表达

5

Tangshan Urban Planning Museum - architectural expression of urban cognition

唐山市城市展览馆
城市认知的建筑表达

唐山市城市展览馆获得2010年美国建筑师学会会刊《建筑实录》杂志"好设计创造好效益"中国最佳公共建筑类杰出大奖,同时被评为年度唯一的"年度最佳项目"。该奖项旨在表彰建筑师和业主之间精诚合作,以优秀的设计来推动并传达社会效应以及扩大公众影响。唐山市城市展览馆受到业内外的认同与社会各界好评,其启示意义不仅在建筑,更在于对自己城市发展观点提供了一个佐证。

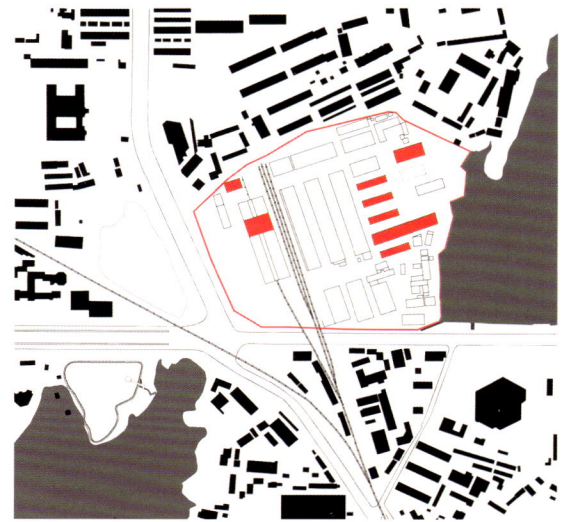

城市心态

因矿而兴的唐山被称为中国近代工业摇篮，这个资源型重工业城市在中国的城市建设史上具有特殊的意义，是大地震后遵循功能主义原则重新规划建设的城市。城市恢复建设的经济制约，以及自然的地震力留给唐山的是规模化、单一化的城市生成模式。三十年的城市重建、振兴、发展，城市更多地致力于未雨绸缪的经济转型、完备基础设施。但在公众心目中，经济发展与城市形象存在巨大反差，相比于其他明星城市，唐山人感到自己城市的乏味。唐山城市的平和与理性遭人诟病，甚至理性得刻板。恰恰相反，与如火如荼的城市相比，唐山的平静与理性却时常让外来者感到差异性的艳羡。

城市认知的建筑表达 —— 另壹种表达

经济的迅速发展让城市充满了欲望的躁动,城市的经济繁荣也带来心态的失衡。没时间来理性思考自己城市的方向。人们只是对城市形象充满期待,急于改变自己。

随着大干快上的城市化热潮蔓延,市场的力量顽强地推动着城市化进程,商业利益最大化的追求与政绩形象目标一拍即合,在种种不切实际的目标驱动下,城市不再坚守理性,不考虑长远的可持续性,也不再遵循发展的客观规律。取代平静和理性的是雄心勃勃突变的城市。人们心中把自己的城市建成大都市景观的渴望,几年就要翻天覆地,心目中追赶所谓的明星城市似乎一蹴而就。要知道罗马不是一天建成的。

人们在对自己城市的一天一天长高而感到欣喜万分的同时,突然发现眼前的高楼大厦似乎与我们日常起居的生活品质无关,我们居住生活的城市没有更舒适,城市形象正取代我们生活中人性的基本要求,重蹈明星城市的覆辙,对未来城市留下巨大隐患,城市的预期将会付出巨大的成本为形象买单。城市生活中拥挤不堪的交通,安全隐患、污染、市政等诸多问题接踵而至。我们只看到了所谓的现代化形象,追求形象的同时却忽略了城市形态与经济发展到一定阶段的必然产物,两者之间存在着必然的逻辑关系。没有人探究支撑摩天楼形象的背后的基础条件是什么。发达健康的城市应是表里如一、内外兼修。城市的心态决定了城市形态,城市的物质形态也决定了城市的性格。

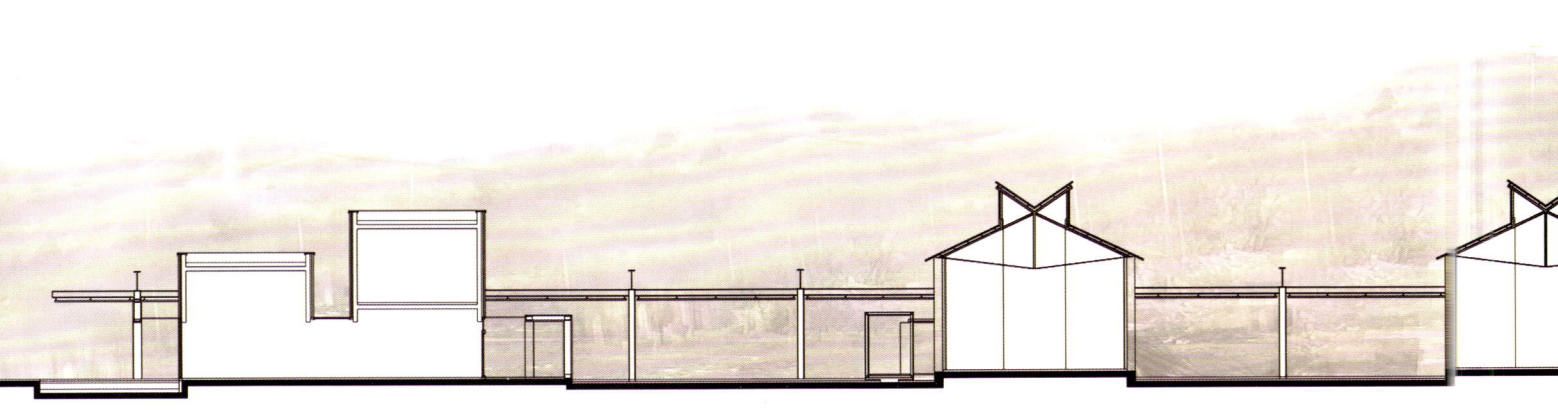

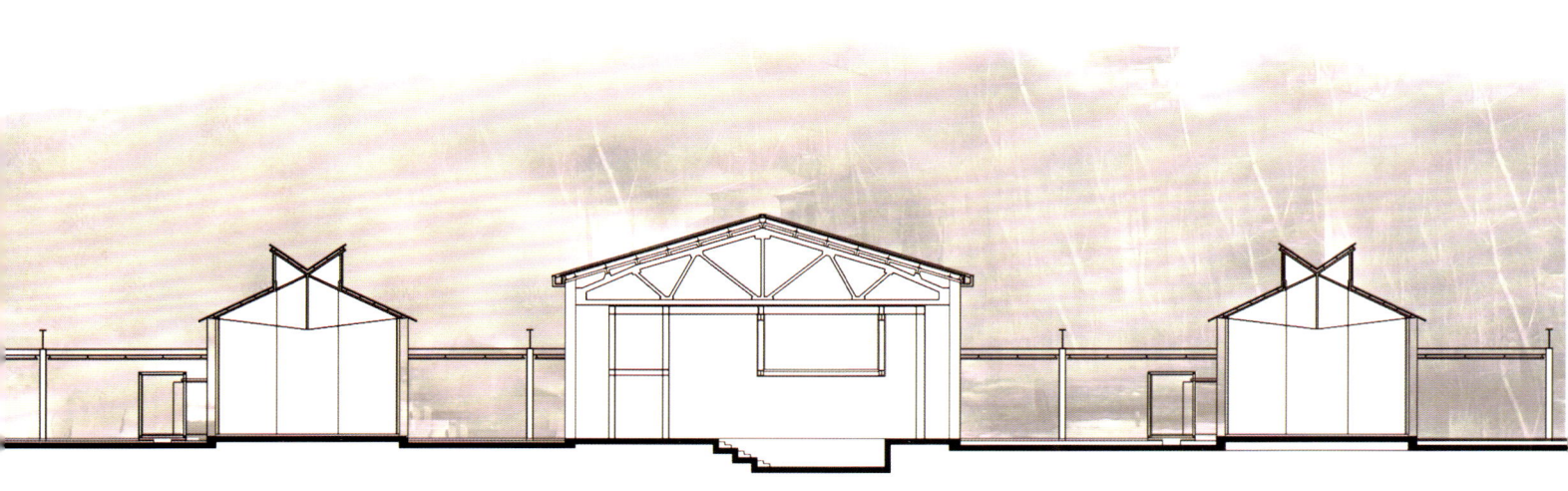

城市认知的建筑表达……另壹种 表达

文化与遗产

文化是一个群体或社会所共有的价值观和社会实践活动中所创造的物质财富与精神财富的总和，以及自身的文化遗存和精神价值，自身创造的一系列文化象征与文化符号。城市是人类文化的集中体现，城市本身就是一个文化积累、积淀的过程，每个城市都有自己独有的文化资源。在当今的中国社会，城市的记忆在大规模、高速度的建设大潮中被无情地抹去，城市记忆的硬盘面临被「格式化」的危险。丧失了记忆的城市变得夸张、浮华、缺少文化和品质。城市肌理由于缺乏了时间的维度和历史的深度而变得单薄脆弱。

城市的更新发展应该是各历史时期文化的沉淀叠加，而不是清零后的重建。急功近利的政绩目标加上资本的无限驱动力，导致在推土机的轰鸣下，大拆促大建，一切都是推倒重来，漠视自身文化的同时，恰恰反映出对自身文化的不自信。

城市在求变、求新、求洋、求大。真正的城市文化和城市特色遭到毁灭性的破坏。有形的物质遗存得不到保护，无形的社会关系、社会秩序需重新建立。市场效应作用下经济力量已使社会分化，房地产按经济原则把城市公民重新划分。若缺少物质和非物质的文化遗产传承的后人了解这段历史，再也看不到清晰而真实的城市年轮，只有靠影像来回顾这座城市中的历史碎片。

城市的发展崇尚利益至上，城市规划的目标宗旨也变成尽可能提高每一个地块的价值，积聚更多的财富。文化搭台、经济唱戏，文化要为经济服务；异化了的文化正在拼命地消灭真正的文化，经济学中的一条规律得到应验："劣币总能驱逐良币"。假冒文化、低俗文化、商业文化占据着城市空间也占据着人的心灵，指导着人们的行动。"我们的文化都被钱收买了，反过来又拿文化来卖钱，这是多么可怕的恶性循环。"试想真实的文化遗迹荡然无存后，取而代之的该是灯红酒绿、光怪陆离的城市万象。

城市策略

城市规划的作用是落实经济社会发展目标，实现经济繁荣；利用技术手段改善人居环境；强化城市的人文关切；把握公平与效率的平衡，平衡和协调利益机制。城市空间不仅有美学意义，同时更应该有价值层面的判断。维护城市的公共利益、长远利益应该是城市规划的核心价值。

唐山这座拥有百年历史的工业城市正应对城市的经济转型，城市规划的策略关注于体现城市特色的公共空间体系建设，以及对城市文化遗产的认知和梳理，对唐山工业遗产等问题进行了定义、普查与评估。对工业遗产的保护与利用提出整体性、策略性方案，其中也包括这一地块前期的分析研究，提出把工业遗址资源和新的城市开放空间资源有机地结合起来，使公共空间体系有了内在的厚度。从文化配套的提升角度来丰富公共空间体系的内涵，以适应城市生活走向多元化和更加开放的市民文化。这是当城市从一个时代步入一个时代时，改头换面最需要做的基础工作。城市展馆的实施完整地体现了这一种工作方法的明确方向。如何能在新旧两个唐山之间搭建起一个记忆的桥梁，让后人还能了解这座城市的来龙去脉？唐山市城市展览馆的项目成为一个例证。

大城山脚下的粮库要被拆除。没有人怀疑拆除这个厂区决定的正确性，因为这个封闭的厂区把位于城市核心的大城山游离到城市生活之外。将之彻底拆除还可增加城市公共空间的容量。粮库虽不具备文物价值和审美价值，始建于日伪时期，经历过大地震而保留下来的建筑，毕竟还有其历史价值，成为屈指可数的城市真实遗迹。城市规划策略认识的明确和坚定的制定为规划的实施提供了保证。前期的研究成果为保留旧建筑实施改造提供了可靠依据，更得益于当时决策者的兼听则明和远见。

城市认知的建筑表达

将工业遗址再利用改造成创意产业空间,渐成一种时尚,我们做的虽也是工业遗址再利用,却无意追逐这种时尚和潮流。把非文物级别的城市历史片段保留下来,用唐山的地方文化、民俗、历史来介绍关于这个城市过去、现在和未来的故事,向市民提供一个完全开放的场所,塑造一个更平民化的环境。这场景不是时尚的道具和布景,而是对现实生活中朴素场景的一种有情调的描述,通过这座建筑的改造实施来唤醒人们对熟视无睹的日常环境的感觉和热情,使之在新的语境中溢彩流光,同时也证明即使如此普通的旧建筑改造也能有品位有分量。这种做法甚至要超过全拆后新建的建筑。城市展馆改造不仅是寓旧于新的再创造,也是我们对城市认知和城市策略的建筑表达,更是表达城市发展理念的一个实践机会。

这个建筑和这座城市有着内在的契合,它们都有凤凰涅槃的意味。不同的是旧建筑改造保留了历史的遗迹,具有了时间的维度,证明了城市的更新没有必要以牺牲其固有性格为代价。城市的魅力来自于平和、内敛、自身的历史文化。

由最初市民化的民俗博物馆设想改为永久的城市规划展馆也有积极意义。展馆本身可以成为城市建设的展品,成为展示城市建设发展的容器,成为了解城市过去、展示城市未来抱负的重要窗口,同时表明规划不仅仅是发展、而且还是保护的理念。这个已被各界广泛认可的改造建筑提醒着城市建设的执行者:城市化不是地震。

城市自信

唐山市城市展览馆的建成,为新城市建设面临如何处理旧建筑遗产问题提供了一个值得借鉴的答案,为如何塑造城市的建筑性格确立了一种方向,为如何优化城市公共空间资源树立了一个范例。其更大的启示意义在于如何能树立城市的自信。震后的规划为唐山现在城区留下了宝贵的遗产:密度低、低容积率高度化一的"水平城市"形态,有别于高楼林立的"垂直城市",道路体系完善、绿化条件几乎在中国主要城市中独一无二,采煤沉陷区的改造所形成的景观已成为城市的名片。这种规划和城市自然山水条件为城市的升华和深化提供了良好的物质条件,建设生态宜居的现代化城市具备得天独厚的自然条件。只有充分利用自身条件,挖掘可利用资源,在解决现实问题的过程中塑造自身特色,才是现实的选择和努力方向。只有根植于自身的历史和文化的城市,才能枝繁叶茂。只有做到文化自觉,才能文化自信、文化自强。

灾后凤凰涅槃的唐山,不可避免地要接受新时代的浴火涅槃。唐山市城市展览馆这个项目可以成为一个例证:城市再生不是脱胎换骨,而是内在气质的放大和自我的觉醒。

城市认知的建筑表达 ——另壹种表达

Tangshan Urban Planning Museum- architectural expression of urban cognition

Tangshan Urban Planning Museum has won 2010 Best Public Project of "Good Design Is Good Business China Awards" awarded by The AIA Journal, Architectural Record. And, it is also honored as the Grand Winner/Project of the Year. This competition in China judges buildings and planning projects on their success in using design to further clients' business or institutional goals, and it has become an important force in raising the quality of design in the world's most dynamic economy. Tangshan Urban Planning Museum is accepted by the industry and outside, and is well received by the public. The revelation of the building not only lies in the architecture, but also provides an evidence for its own urban development viewpoints.

Urban Attitude

Booming from rich mining resource, Tangshan is regarded as the cradle of China's modern industry. The resource-based heavy industrial city has a very unique meaning in the history of China's urban construction. It had been replanned and reconstructed, conforming to functionalism principle, after the major earthquake. The economic constraints in the post-earthquake recovery and the risks of natural earthquake forces led to a large-scale, unitary urban generation model for Tangshan. During three decades of urban renewal, revitalization, and development, the city is more committed to the precautionary urban economic restructuring plan and complete infrastructure. But in the public mind, there is a striking contrast between economic development and urban image. Compared with other star cities, people feel their own city is boring. Tangshan city's calm and rationalism is criticized, and somehow too rigid. On the other hand, instead of cities in full swing, Tangshan's calm and rationalism are often envied by outsiders.

The rapid economic development filled the cities with restlessness of desire. The cities' economic prosperity has also brought the imbalance of the mentality and they have no time to rationally think about their own urban development. People are only looking forward to the city image and eager to change themselves. With a big wave of rapid urbanization, market forces tenaciously promote the urbanization process, and the pursuits of maximizing commercial benefits and of official achievement images chime in easily. Driven by all kinds of unrealistic goals, the city is no longer adhering to rationalism. It does not consider long-term sustainability, and no longer follow the objective law of development. Calm and rationalism is replaced by ambitious mutant city. People are filled with desire to build their own city to a metropolitan. Cities are constructed to be revolutionarily different in a few years. Aimlessly chasing after so-called star cities seems to reach your aim in one move, but Rome is not built in one day.

When people are delighted at the gradually growing-up cities, all of a sudden, they find out that it seems skyscrapers have nothing to do with the quality of our everyday life. The cities we are living in haven't become more comfortable therefore. The urban image is taking place of the fundamental necessities of our daily life. Heavy traffic, potential safety hazards, pollution, civil engineering and related problems successively occur in the city life. Repeating the mistakes of star cities will leave huge risk for the future city, and pay huge costs for the expected city image. What we see is only so-called modern image. But we ignore the natural product of urban form and economic development to a certain stage while pursuing appearance. There is inevitable logic relation between the two. Nobody explores the basis behind the image of skyscrapers. The healthy and developed cities should be integrated and cultivated internally and externally. The urban attitude determines the urban form; the urban physical form settles the urban characteristics.

Culture and Heritage
Culture is the common value cherished by a group or a society, the sum of material wealth and spiritual wealth created during social practice, as well as, a series of self-created symbols and signals embodying their own culture heritages and spiritual values. City is a concentrated expression of human culture; the city itself is a cultural accumulation process. Each city has its own specific and unique culture resource. In today's Chinese society, the city's memory is inexorably erased in large-scale, high-speed construction tide; the hard disk of urban memories is running the risk of "being formatted". Cities suffered memory loss become exaggerate, flashy and lack of culture and quality. Urban fabric turns thin and fragile due to lack of time dimension and historical depth.

Urban renovation development should be the deposit and superimposition of every historical period rather than reconstruction after all clear. Political achievements goal of eager for quick success and instant benefit with unlimited driving force of capital, result in everything being pushed down and started all over again under the bulldozer's roar, large-scale demolition to promote construction. Ignoring their own culture just reflects their uncertainty.

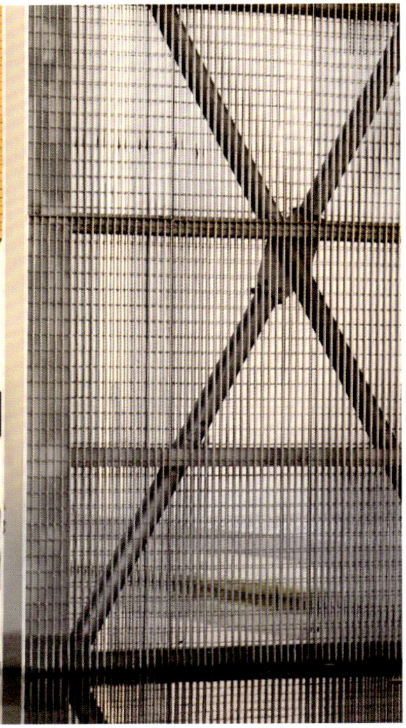

Cities are seeking change, novelty, foreignness, and largeness. True urban culture and urban characteristics have suffered from devastating damage. Corporeal material remains couldn't be protected; invisible social relations and social order need to be re-established. Economic power under market effects has disintegrated the society; real estate has re-classified urban citizens according to economic principles. If future generations who are lack of tangible and intangible culture heritage get to know this period of history, they can no longer clearly see the authentic historical layers of a city, but only rely on the image to recall the historical fragments of the city.

Economic interests are the ultimate aim of urban development, therefore urban planning's purposes have also become trying to maximize the value of each block and accumulate more wealth. Culture sets up the stage, economics perform, and that is to say culture should serve for economics. The distorted culture is trying hard to eliminate the real culture, an economics law is fulfilled: "Bad money drives out good." Counterfeit culture, vulgar culture, and commercial culture take up urban space, and also occupy the human mind, guide people's actions. Our culture is bought by money, and in turn, culture is used to make money. What a frightening vicious cycle! Imagine the real culture relics are gone, debauchery and bizarre city takes place.

Urban Strategies

Urban planning is to implement economic and social development goals, and achieve economic prosperity; by use of technology to improve the living environment; to realize the city's cultural concerns; to keep the balance between fairness and efficiency; to develop the mechanisms of balancing and coordinating interests. Urban space not only has its aesthetic significance, but also should have judgment on value. To maintain the city's public interest and long-term interest should be the core values of urban planning.

Tangshan, an industrial city with over 100 years of history, is dealing with the city's economic restructuring. Its urban planning strategies focus on the public space system development reflecting the city's characteristics, awareness and sorting of urban cultural heritage. It defines, surveys and evaluates the Tangshan industrial heritage and other issues, proposes integral strategic plan of industrial heritage's preservation and use. This plot's early phase analysis and research is also included. It proposed to organically combine industrial relics resource and new open urban space resource, to enable public space system to have inherent thickness. To enrich the connotation of public space system from the perspective of culture support, it's crucial to get accustomed to city life heading for diversification and a more open folk culture. This is the basic work to achieve a face lift when the city enters into a new era from an old one. The implementation of the Urban Planning Museum completely reflects the clear direction of this working method. How to build up a mnemonic bridge between the old and new Tangshan cities, so future generations can understand the context of the city? Tangshan Urban Planning Museum project turns out to be a demonstration.

The granary at the foot of Dacheng Mountain has to be removed. No one doubts the correctness of the decision to demolish it, because this enclosed factory isolates Dacheng Mountain located at the city center to be outside the urban life. Thoroughly removal will also increase the capacity of urban public space. Although the granary does not bear heritage or aesthetic value, after all it still has its historical value since it was first built in Puppet period, then experienced the earthquake and was preserved. It's one of the few true monuments of the city. Clear understanding and decisive execution of urban planning strategies provide guarantee for the implementation of planning. Preliminary research results supply reliable evidence for preserved building renovation. And, it is more due to policy-makers' hearing all parties and visions.

Architectural Expression of Urban Cognition

Creating an urban planning museum and park out of an old granary complex re-uses the industrial site and transforms it into a creative industrial place, and this has become a metropolitan fashion. Although this project is an example of regenerating industrial sites, it has no intention to chase this fashion and trend. Keep the non-preserved fragment of the city history, and tell the stories of this city's past, present and future with Tangshan local culture, folklore and history. Providing a completely open public place and creating a more civilian-oriented environment, this scene is not a fashion props and scenery, but a romantic description of the simple scene of real life scenario. Through the implementation of the building renovation, the architect means to wake people's feeling and passion about the daily environment which they used to turn a blind eye to, and make it shiny in the new context. Meanwhile, it proves that even plain old building like this could be renovated to be classy and authoritative. This approach will even surpass the newly built buildings after demolition.

The building and this city are inter-dependent both are featured with nirvana significance. The difference lies in that the old building renovation keeps historical relics and has a time dimension, which proves that the city's update does not have to pay the price of sacrificing its own character. The city's charm comes from the calm, restrained, and inherent historical culture.

The urban planning museum renovation is not only a re-creation blending the old with the new, but also architectural expression of our urban comprehension and urban policy, and moreover a practical opportunity of expressing urban development concept. It also has a positive meaning that the first civilian "folk museum" idea was replaced by decision-makers with "Urban Planning Museum". The museum itself could be the exhibit of urban construction, the container of displaying urban construction development, and the window into the past to understand the city and displaying the city's future vision. At the same time, it shows that planning is not just development, but also idea of preservation. This widely accepted renovated building reminds the executer of urban construction: urbanization is not earthquake.

西立面图

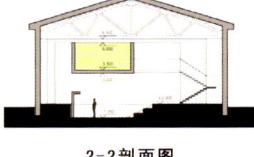
2-2剖面图

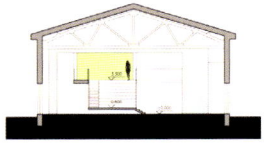
3-3剖面图

Urban Confidence

Tangshan Urban Planning Museum provides an answer to the question of how the new urban construction deals with the remaining architectural heritages; chooses a direction of how to form the urban architectural characters; sets up an example of how to create some high-quality public space resource. The most important is that it enlightens us how to build urban confidence. The post-earthquake planning of Tangshan leaves precious heritage in present urban area: low density, low floor area ratio (FAR), extremely uniform "horizontal city" form different from "vertical city" packed with high-rise buildings, the well-organized road and green systems almost distinguished from other major cities in China, and landscaped formed by renovation of sinking land in coalmining areas having become the city's name card. The urban planning and natural landscape have provided a good physical condition for upgrading and improving the quality of this city. Tangshan has a unique natural condition for building an ecologically livable modern city. Only making full use of its own conditions, exploring available resource, shaping its own characteristics in the process of solving practical problems are the practical choices and development directions. Only rooted in its own history and culture, could the city flourish. We can only be culturally confident and self-reliance by being cultural awareness.

The nirvana of the post-earthquake Tangshan is inevitable of the test of modern times. Tangshan Urban Planning Museum is able to prove: the reborn of a city is not a complete change, but magnification of inner temperament and awareness of ego.

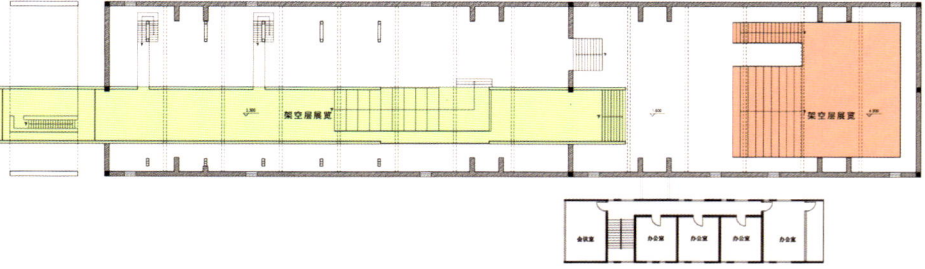

2010 | China Awards

good design is good business

PROJECT   Tangshan Urban Planning Museum and Park

ARCHITECT   Urbanus Architecture and Design

CLIENT   Tangshan Urban Planning Administration Bureau

PROJECT OF THE YEAR

ARCHITECTURAL RECORD

唐山城市规划展览馆及公园

2010年获得美国建筑师学会会刊《建筑实录》（*Architectural Record*）杂志及其母公司麦格劳-希尔建筑信息公司第三届好设计创造好效益中国2010年"最佳公共建筑类杰出大奖"（2010 China Award , Good Design Is Good Business）。此奖项为《商业周刊》/《建筑实录》中国设计大奖，每两年颁发一次，旨在表彰建筑师和业主之间精诚合作，以优秀的设计来推动商业目的，传达社会效应以及扩大公众影响。唐山市城市规划展览馆是当年度唯一的大奖项目。

2008年获《世界建筑》杂志社主办第四届WA奖佳作奖（2008 WA Chinese Architecture Award Honorable Mention）

2011年获《城市·环境·设计》杂志社主办 UED博物馆建筑设计奖（2011 UED museum design award）

潘家峪惨案纪念馆

设计/建成：1997/1998

建筑面积：1260平方米

方案：沈瑾

施工图设计：许智梅

建设地点：河北省唐山市丰润区

建设单位：河北省唐山市丰润区宣传部

井陉矿万人坑纪念馆

设计/建成：2000/2007

建筑面积：1460平方米

方案：沈瑾

施工图设计：汪会东、高建成

景观设计：房木生

顾问建筑师：李拱臣

雕塑浮雕：李林琢

建设地点：河北省石家庄市井陉矿区

建设单位：河北省石家庄市井陉矿区宣传部

看得见风景的房间

设计/建成：2004/2004

建筑面积：1286平方米

方案：沈瑾

施工图设计：汪会东、王卫国

室内设计：张晔

建设地点：河北省唐山市凤凰山公园

建设单位：河北省唐山市人民防空办公室

唐山南湖展馆

设计/建成：2005/2006

建筑面积：960平方米

方案：沈瑾

施工图设计：李强

室内设计：张晔

建设地点：河北省唐山市南湖公园

建设单位：唐山市园林局

唐山城市规划展览馆

设计/建成：2006/2008

建筑面积：5650平方米

项目策划主持：沈瑾

建筑师：王辉、刘旭

建筑/景观/室内设计：URBANUS都市实践（北京）

展陈设计：天津奥林华亚展示有限公司

建设地点：唐山市大城山博物馆公园

建设单位：唐山市城乡规划局

沈 瑾 建 筑 设 计 作 品 集

顾　　问：黄为隽
艺术总监：程大鹏
平面设计：孔煜
责任编辑：王忠波
中文文字：沈瑾
英文翻译：孙炼
英文校对：David Cobb
图片摄影：沈瑾　张广源　杨超英　杨滔

另壹种 表达 IFFERENT XPRESSION

图书在版编目(CIP)数据

另一种表达 / 沈瑾著.
—北京：中央编译出版社,2012.7
ISBN 978-7-5117-1421-3

Ⅰ.①另…
Ⅱ.①沈…
Ⅲ.①建筑设计—作品集—中国—现代 Ⅳ.①TU206
中国版本图书馆CIP数据核字(2012)第130334号

另一种表达

责任编辑	王忠波
责任印制	尹 珺
装帧设计	孔 煜
出版发行	中央编译出版社
地　　址	北京西城区车公庄大街乙5号鸿儒大厦B座(100044)
电　　话	(010)52612345(总编室)　(010)52612339(编辑室)
	(010)66161011(团购部)　(010)52612332(网络销售)
	(010)66130345(发行部)　(010)66509618(读者服务部)
网　　址	www.cctphome.com
经　　销	全国新华书店
印　　刷	北京顶佳世纪印刷有限公司
开　　本	787毫米×960毫米　1/12
字　　数	145千字
印　　张	12.5
版　　次	2012年7月第1版第1次印刷
定　　价	158.00元